UNIT 1

1.1 Sequences

Page 1 ***HWK 1M***

1. (a) 29, 36, 43, 50, 57, 64 (b) 29, 19, 9, –1, –11, –21 (c) 29, 290, 2900, 29000, 290000, 2900000

2. (a) subtract 0.2 (b) multiply by 2 (c) divide by 10 (d) add 2

3. (a) 3, 12, 48, 192, 768 (b) –4, –1, 2, 5, 8, 11 (c) 26, 21, 16, 11, 6, 1

4. (a) Ⓔ Ⓔ Ⓔ Ⓔ Ⓔ Ⓔ Ⓔ Ⓔ Ⓔ (b) 49

5. Luke doubles each time, Ali adds 1 then 2 then 3

Page 2 ***HWK 1E***

1. (a) 4, 7, 13, 25 (b) 4, 13, 40, 121 (c) 4, 12, 28, 60 **2.** (a) 6, 10, 14, 18, 22, 26
(b) 30, 23, 16, 9, 2, –5 (c) 57, 49, 41, 33, 25, 17 **3.** 11 **4.** 25

5. 4, 10, 22, 46 **6.** multiply by 3 and subtract 1

1.2 Fractions

Page 2 ***HWK 1M***

1. (a) $2\frac{3}{4} = \frac{11}{4}$ (b) $3\frac{2}{5} = \frac{17}{5}$ (c) $4\frac{5}{6} = \frac{29}{6}$ (d) $2\frac{2}{5} = \frac{12}{5}$

2. (a) $3\frac{1}{2}$ (b) $4\frac{1}{3}$ (c) $2\frac{4}{5}$ (d) $3\frac{1}{4}$ (e) $3\frac{2}{7}$
(f) $1\frac{5}{6}$ (g) $3\frac{3}{10}$ (h) $2\frac{5}{8}$ (i) $2\frac{1}{7}$ (j) $5\frac{4}{5}$

Page 3 ***HWK 1E***

1. (a) $\frac{14}{3}$ (b) $\frac{9}{4}$ (c) $\frac{23}{4}$ (d) $\frac{15}{2}$ (e) $\frac{19}{5}$ (f) $\frac{43}{8}$ (g) $\frac{25}{9}$
(h) $\frac{37}{6}$ (i) $\frac{22}{5}$ (j) $\frac{57}{10}$ **2.** $\frac{33}{8} \to 4\frac{1}{8}$, $\frac{17}{5} \to 3\frac{2}{5}$, $\frac{29}{8} \to 3\frac{5}{8}$, $\frac{23}{6} \to 3\frac{5}{6}$, $\frac{19}{8} \to 2\frac{3}{8}$

Page 3 ***HWK 2M***

1. $\frac{25}{30}$ **2.** (a) $\frac{6}{8} - \frac{1}{8} = \frac{5}{8}$ (b) $\frac{5}{15} + \frac{2}{15} = \frac{7}{15}$ (c) $\frac{11}{20} - \frac{5}{20} = \frac{6}{20} = \frac{3}{10}$ **3.** (a) $\frac{5}{16}$
(b) $\frac{4}{20} = \frac{1}{5}$ (c) $\frac{6}{30} = \frac{1}{5}$ (d) $\frac{9}{12} = \frac{3}{4}$ (e) $\frac{2}{21}$ (f) $\frac{7}{40}$ **4.** $\frac{1}{8}$

5. (a) true (b) false (c) true (d) true (e) false (f) true

Page 4 ***HWK 2E***

1. (a) $\frac{4}{15}$ (b) $\frac{11}{12}$ (c) $\frac{31}{40}$ (d) $\frac{33}{56}$ (e) $\frac{1}{14}$ (f) $\frac{47}{90}$
(g) $\frac{5}{24}$ (h) $\frac{13}{70}$ **2.** $\frac{6}{35}$ **3.** (a) $2\frac{7}{12}$ (b) $5\frac{5}{6}$ (c) $1\frac{5}{12}$
(d) $1\frac{17}{24}$ (e) $4\frac{1}{6}$ (f) $1\frac{5}{8}$ **4.** (a) $\frac{13}{60}$ (b) $\frac{7}{15}$

Page 5 ***HWK 3M***

1. (a) 18 (b) 24 (c) 56 (d) 25 (e) 18 (f) 15 (g) 21 (h) 24

2. 12 **3.** 900 m **4.** B **5.** 4 km **6.** (a) $\frac{\boxed{2}}{3}$ (b) $\frac{\boxed{7}}{8}$ (c) $\frac{5}{\boxed{6}}$

Page 5 ***HWK 3E***

1. (a) $\frac{1}{10}$ (b) $\frac{4}{7}$ (c) $\frac{1}{4}$ (d) $\frac{1}{14}$ (e) $\frac{1}{2}$ (f) $\frac{3}{8}$ (g) $\frac{10}{21}$ (h) $\frac{7}{22}$

2. B by $\frac{1}{40}$ m^2 **3.** (a) false (b) true (c) true (d) false **4.** $2\frac{3}{4}$ m^2

1.3 Properties of numbers

Page 6 ***HWK 1M***

1. 6 **2.** 11, 13, 17, 19 **3.** 35, 77, 21, 42 **4.** True **5.** 27

6. 11, 17 **7.** 1, 32, 2, 16, 4, 8 **8.** 2, 4, 10 or 1, 5, 10

Page 7 ***HWK 1E***

1. (a) 1, 30, 2, 15, 3, 10, 5, 6 (b) 1, 48, 2, 24, 3, 16, 4, 12, 6, 8 (c) 6

2. 1, 3, 6, 10, 15, 21, 28, 36, 45 **3.** (a) 15, 30, 45, 60, 75, 90 (b) 20, 40, 60, 80, 100, 120

(c) 60 **4.** 17 **5.** 90 **6.** 7 **7.** divisible by 10 **8.** every 35 days

Page 8 ***HWK 2M***

1.

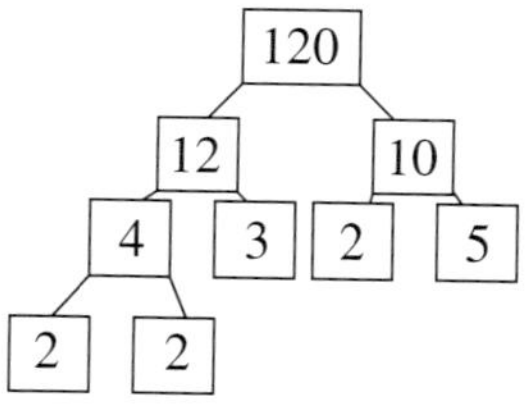

$120 = 2 \times 2 \times 2 \times 3 \times 5$

2. (a) $50 = 2 \times 5 \times 5$

(b) $140 = 2 \times 2 \times 5 \times 7$

(c) $240 = 2 \times 2 \times 2 \times 2 \times 3 \times 5$

(d) $420 = 2 \times 2 \times 3 \times 5 \times 7$

3. (a) $630 = 2 \times 3 \times 3 \times 5 \times 7$ (b) $1560 = 2 \times 2 \times 2 \times 3 \times 5 \times 13$ (c) 30

Page 8 ***HWK 2E***

1. (a) 8 (b) 8 (c) 27 (d) 2 (e) 6 (f) 512 (g) 9

(h) 3 (i) 8000 (j) 1 **2.** 4^5 **3.** (b), (c) and (d) **4.** 2^6

5. True **6.** (a) 144 (b) $\frac{1}{16}$ (c) 10000 (d) 0.16 (e) 0.001

1.4 Negative Numbers

Page 9 ***HWK 1M***

1. (a) true (b) false (c) false (d) true (e) true (f) false

2. 4, –4, –2, –5, –4 **3.** (a) –5 (b) –8 (c) –1 (d) 4 (e) –5

(f) –9 (g) –5 (h) 0 **4.** A **5.** (a) – 3 – 1 = –4

(b) 6 – 9 = –3 (c) –7 + 8 = 1 **6.** –1

Page 10 ***HWK 1E***

1. (a) true (b) false (c) false (d) false (e) true (f) true **2.** 7, –2, 9, –4, 5

3. (a) 5 (b) 1 (c) –2 (d) 9 (e) 5 (f) 1 (g) –1

(h) –6 **4.** 20 **5.** –6 **6.** (a) –2 – (–3) = 1 (b) 5 – (–1) = 6

(c) –3 + (–4) = –7 **7.** A **8.**

–2	–3	2
3	–1	–5
–4	1	0

Page 11 ***HWK 2M***

1. (a) –12 (b) 12 (c) 24 (d) –14 (e) 10 (f) –18 (g) –36

(h) 4 (i) 42 (j) 27 (k) –40 (l) –60 **2.** –24°C **3.** –4 × (–4) by 36

4. 3, –9, –18, 90 **5.** (a) –6 × (5) = –30 (b) 9 × (–4) = –36 (c) –10 × (–7) = 70

(d) (7) × (–2) = –14 (e) –8 × (–4) = 32 (f) (–6) × (–8) = 48

Page 11 ***HWK 2E***

1. (a) –4 (b) 4 (c) 4 (d) –8 (e) –4 (f) 5 (g) –7 (h) –2

(i) –12 (j) –4 (k) –7 (l) –5 **2.** –8, –2 **3.** (a) false (b) true

(c) true (d) true (e) false (f) true **4.** –60, 30, –6, –2

5.

×	–5	3	–2
–3	15	–9	6
4	–20	12	–8
–7	35	–21	14

1.5 Area and Perimeter

Page 12 ***HWK 1M***

1. (a) 45 cm^2 (b) 16 cm^2 (c) 30 cm^2 (d) 24 cm^2 (e) 42 cm^2 (f) 24 cm^2

2. (a) 48 m^2 (b) 12 m^2 (c) 60 m^2 (d) 60 m^2 **3.** (a) 120 cm^2 (b) 84 cm^2 (c) 92 cm^2 **4.** 80 cm^2 **5.** 2×9 rectangle (perimeter 22 cm) and 1×18 rectangle (perimeter 38 cm)

Page 14 ***HWK 1E***

1. shape A by 64 cm^2 **2.** 35 cm **3.** 40 m^2 **4.** 25 m **5.** (a) 100 m^2 (b) £1585

Page 15 ***HWK 2M***

1. (a) 65 cm^2 (b) 120 cm^2 (c) 170 cm^2 (d) 117 cm^2 (e) 160 cm^2 (f) 240 cm^2 **2.** (a) 7 m (b) 9 m **3.** shape A by 5 cm^2

Page 16 ***HWK 2E***

1. (a) 1500 (b) £591 (c) £29.55 **2.** 4 cm

3. 163 m^2 **4.** 30 cm **5.** 72 cm **6.** 4

UNIT 2

2.1 Rounding off and estimating

Page 18 ***HWK 1M***

1. (a) 8.7 (b) 6.5 (c) 7.1 (d) 4.1 (e) 5.2 (f) 10.8 (g) 0.1 (h) 3.0
2. 6.782 **3.** (a) 2.8 (b) 1.3 (c) 5.1 (d) 11.6 **4.** (a) 7.81 (b) 72.86 (c) 4.75
(d) 4.85 (e) 1.86 (f) 208.67 (g) 9.27 (h) 133.32

Page 18 ***HWK 1E***

1. (a) 3.48 (b) 3.36 (c) 1.13 (d) 19.90 (e) 11.45 (f) 1.18 (g) 6.00 (h) 0.46
2. 8.135 **3.** 3 **4.** 9.25 **5.** 2.375

Page 19 ***HWK 2M***

1. 800 **2.** 800 **3.** 5600 **4.** 5600 **5.** 450 **6.** (a) B (b) A (c) B
(d) B (e) A (f) C (g) C (h) A (i) A (j) C (k) A
(l) B (m) C (n) C (o) A

Page 20 ***HWK 2E***

1. ≈£50 **2.** ≈9000 g (9 kg) **3.** ≈£1500 **4.** (a) $20.6 \div 5 = 4.12$ (b) $49 \times 20.2 = 898.8$
(c) $8.1 \times 32 = 259.2$ (d) $42 \times 6.8 = 285.6$ (e) $2.8 + 13.9 = 16.7$ (f) $3012 \div 4.8 = 627.5$
5. ≈800 m **6.** ≈£190 **7.** ≈£1

2.2 Using Algebra

Page 20 ***HWK 1M***

1. (a) false (b) true (c) true (d) false **2.** $5a + 4b$ **3.** $4m + 7n$ **4.** $3a + 6$ **5.** $2x + 2$
6. $5w + 3$ **7.** $2m + 4n$ **8.** $4 + p$ **9.** $3x + 1$ **10.** $2a + 3$ **11.** A – E, B – D **12.** $10x + 2y + 5$

Page 21 ***HWK 1E***

1. (a) false (b) true (c) false (d) true (e) true (f) true (g) false (h) true (i) true
2. (a) $\boxed{4a - a}$ (b) $\boxed{5 \div a}$ (c) $\boxed{2a + a + a}$ and $\boxed{4a}$ (d) $\boxed{a \div 2}$ (e) 4 **3.** £$(3n + 4)$ **4.** $(5w + 1)$ metres

Page 22 ***HWK 2M***

1. (a) $4(3n + 6)$ (b) $8(5n - 3)$ (c) $3(n + 4) + 1$ (d) $9(n - 2) - 4$ (e) $5(n + 5)^2$ (f) $7(n^2 + 2)$
2. (a) $n \to \boxed{\times 4} \to \boxed{+1} \to \boxed{\times 6} \to$ (b) $n \to \boxed{\times 5} \to \boxed{-8} \to \boxed{\times 4} \to$
(c) $n \to \boxed{\text{square}} \to \boxed{+4} \to \boxed{\times 5} \to$ (d) $n \to \boxed{\times 3} \to \boxed{-6} \to \boxed{\div 7} \to$
(e) $n \to \boxed{-4} \to \boxed{\text{square}} \to \boxed{\times 7} \to$ (f) $n \to \boxed{+5} \to \boxed{\text{square}} \to \boxed{\times 8} \to$
3. $50 - m$ **4.** $35 + x - y$ **5.** $x - y - 7$ **6.** $7y$

Page 22 ***HWK 2E***

1. (a) $n-5$ (b) $m-4$ (c) $m+n-9$ (d) $m-3$ **2.** $3n-15$

3. (a)
- $2a+2b$
- $a+b$ | $a+b$

(b)
- $3m+3n$
- $2m$ | $m+3n$
- $m-n$ | $m+n$ | $2n$

(c)
- $4a$
- $2a+b$ | $2a-b$
- $a+2b$ | $a-b$ | a

(d)
- $5a+2b$
- $3a$ | $2a+2b$
- $2a+b$ | $a-b$ | $a+3b$

(e)
- $2a+4b$
- $a+4b$ | a
- $a+3b$ | b | $a-b$

(f)
- $4c+d$
- $3c+d$ | c
- $3c+d$ | $2d$ | $c-2d$

4. £$(3n+7)$

2.3 Fractions, decimals, percentages

Page 23 ***HWK 1M***

1. (a) true (b) false (c) true (d) false (e) true (f) true **2.** (a) $\frac{1}{5}$ (b) $\frac{9}{100}$ (c) $\frac{9}{25}$ (d) $\frac{3}{4}$ (e) $\frac{7}{1000}$ (f) $\frac{1}{40}$ (g) $\frac{73}{100}$ (h) $\frac{1}{125}$ (i) $\frac{9}{20}$ (j) $\frac{4}{25}$ **3.** (a) 0.4 (b) 0.55 (c) 0.38 (d) 0.52 (e) 0.125

4. (a) 0.8 (b) $\frac{19}{20}$ (c) 0.17 (d) $\frac{7}{25}$ (e) $\frac{3}{50}$ (f) 0.49

Page 24 ***HWK 1E***

1. $0.\dot{1}$ **2.** $0.\dot{2}8571\dot{4}$ **3.** $0.\dot{3}$ **4.** $0.\dot{5}7142\dot{8}$ **5.** $0.\dot{0}\dot{9}$ **6.** $0.\dot{8}5714\dot{2}$ **7.** $0.\dot{8}$ **8.** $0.\dot{1}\dot{2}$ **9.** $0.\dot{3}\dot{6}$ **10.** $0.\dot{0}\dot{2}$

Page 24 ***HWK 2M***

1. $\frac{7}{20} = 35\%$, $\frac{6}{25} = 24\%$, $\frac{9}{40} = 22.5\%$, $\frac{1}{4} = 25\%$: $\frac{9}{40}, \frac{6}{25}, \frac{1}{4}, \frac{7}{20}$ **2.** (a) $\frac{41}{100}$ (b) $\frac{6}{25}$ (c) $\frac{3}{50}$ (d) $\frac{9}{20}$ (e) $\frac{8}{25}$ **3.** (a) 0.03 (b) 0.3 (c) 0.16 (d) 0.75 (e) 0.85 **4.** (a) 15% (b) 27.5% (c) 2.5% (d) 12.5%

5. (a) 33% (b) 64% (c) 9% (d) 14% (e) 130% **6.** $\frac{16}{25}$

Page 25 ***HWK 2E***

($\frac{12}{25}, \frac{36}{75}$, 0.48, 48%), ($\frac{13}{50}, \frac{39}{150}$, 0.26, 26%), ($\frac{4}{5}, \frac{16}{20}$, 0.8, 80%), ($\frac{52}{80}, \frac{13}{20}$, 0.65, 65%), ($\frac{2}{25}, \frac{12}{150}$, 0.08, 8%), ($\frac{3}{8}, \frac{15}{40}$, 0.375, 37.5%), ($\frac{3}{4}, \frac{21}{28}$, 0.75, 75%)

2.4 Geometrical Reasoning

Page 25 ***HWK 1M***

1. $a = 69°$ **2.** $b = 45°$, $c = 45°$, $d = 135°$ **3.** $e = 127°$, $f = 127°$, $g = 127°$ **4.** $h = 47°$

5. $i = 78°$, $j = 68°$, $k = 112°$ **6.** $l = 32°$ **7.** $m = 34°$, $n = 34°$ **8.** $p = 69°$ **9.** $q = 32°$, $r = 123°$

10. $s = 56°$, $t = 68°$ **11.** $u = 73°$, $v = 39°$ **12.** $w = 75°$, $x = 75°$, $y = 75°$

Page 26 ***HWK 1E***

1. $A\hat{D}B = A\hat{B}D$, $B\hat{D}C = D\hat{B}C$, $A\hat{D}C = A\hat{D}B + B\hat{D}C = A\hat{B}D + D\hat{B}C = A\hat{B}C$

2. $a+b+c = 180°$, $d+e+f = 180°$, $g+h+i = 180°$, $a+b+c+d+e+f+g+h+i = 540°$

3. Use alternate angles

Page 27 ***HWK 2M***

1. $a = 69°$ **2.** $b = 68°$ **3.** $c = 84°, d = 96°, e = 63°$ **4.** $f = 77°$ **5.** $g = 115°, h = 89°$
6. $i = 83°$ **7.** $j = 114°$ **8.** $k = 67°$ **9.** (a) 84° (b) 96° (c) 115° (d) 65°

Page 27 ***HWK 2E***

1. $a = 78°$ **2.** $b = 120°$ **3.** $c = 71°, d = 109°$ **4.** $e = 134°, f = 134°, g = 46°$
5. $h = 97°, i = 83°, j = 83°, k = 116°$ **6.** $l = 50°$ **7.** $m = 63°, n = 35°$ **8.** $p = 66°, q = 71°$
9. $r = 45°$ **10.** $s = 102°, t = 49°$ **11.** $u = 70°, v = 110°, w = 69°$ **12.** $x = 60°$

2.5 Construction and Locus

Page 28 ***HWK 1M***

1. 58° **2.** (a) 5.8 cm (b) 7.2 cm (c) 10.7 cm **3.** 49° **4.** 66°

Page 29 ***HWK 1E***

1. circle of radius 6 cm **2.** bisector of $B\hat{A}C$

3. 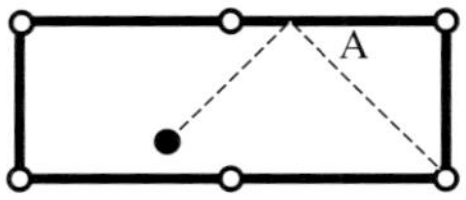

5. 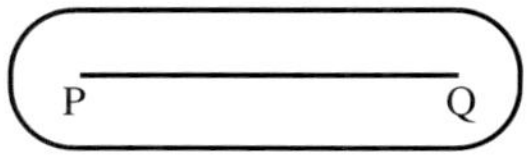

6. Any reasonable attempt, eg. 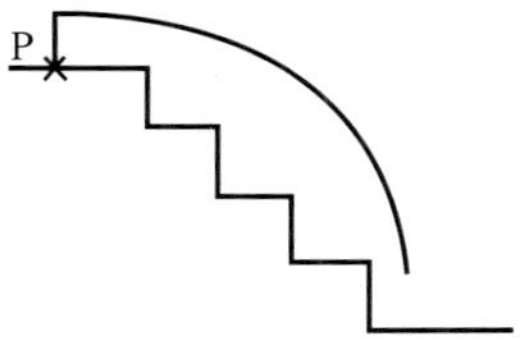

Page 30 ***HWK 2M/2E***

3. (e) 135° (f) 135° **5.** (d) $Q\hat{P}X \approx 20°$, $P\hat{R}Y \approx 32°$

2.6 Circles

Page 31 ***HWK 1M***

1. (a) 9 cm (b) 18 cm **2.** (a) 4 m (b) 8 m **3.** (a) 6 m (b) 12 m
4. (a) 6.5 mm (b) 13 mm **5.** (a) 14.5 cm (b) 29 cm **6.** (a) 5.5 m (b) 11 m
7. (a) 16 m (b) 32 m **8.** (a) 24 mm (b) 48 mm **9.** (a) 9 cm (b) 18 cm
10. (a) 10.5 m (b) 21 m

Page 31 ***HWK 2M***

1. (a) 25.1 cm (b) 47.1 cm (c) 18.8 m (d) 44.0 m **2.** 94.2 m **3.** 50.3 mm
4. Anna by 11.3 m **5.** 21.4 m

Page 32 ***HWK 2E***

1. 36.0 cm **2.** 17.9 cm **3.** 22.5 cm **4.** shape A by 3.4 cm **5.** 37.6 cm

Page 33 ***HWK 3M***

1. 28.3 m^2 **2.** 314.2 mm^2 **3.** 132.7 cm^2 **4.** 1452.2 cm^2 **5.** 153.9 mm^2 **6.** 12.6 m^2
7. 962.1 cm^2 **8.** 254.5 m^2 **9.** 1385.4 cm^2 **10.** 380.1 cm^2 **11.** shape B by 1.98 cm^2 **12.** 61.8 cm^2

Page 33 ***HWK 3E***

1. 12.6 cm^2 **2.** 157.1 cm^2 **3.** 427.6 cm^2 **4.** 2748.9 m^2 **5.** 81.1 m^2 **6.** 553.1 cm^2

UNIT 3

3.1 Written Calculations

Page 35 ***HWK 1M***

[1] 2	5	6	[2] 2		[3] 9	3	[4] 5
1			[5] 4	0	8		4
[6] 8	[7] 4	0	0		[8] 7	5	0
[9] 6	6		[10] 8	[11] 4			0
	[12] 8	5		[13] 3	4	[14] 6	
[15] 7	0		[16] 2	9		9	
2			[17] 1	6		3	
[18] 8	3	7	4		[19] 8	0	5

Page 35 ***HWK 1E***

1. 6.8 **2.** 930 **3.** 169 **4.** 0.108 **5.** 0.94 **6.** 40.6 **7.** 0.26
8. 0.056 **9.** 5800 **10.** 0.747 **11.** 52.2 **12.** 0.046 **13.** £5.66 **14.** 12
15. 6.2 ÷ 100 larger by 0.001 **16.** 'Calypso' by 0.1 p **17.** 11.77 kg **18.** £2.76

Page 36 ***HWK 2M***

1. a = 7.3 m, b = 7.8 m **2.** a = 13.2 kg, b = 13.6 kg **3.** a = 64 m, b = 72 m
4. a = 0.44 cm, b = 0.48 cm **5.** a = 40 g, b = 55 g **6.** a = 5.5 m, b = 7.5 m
7. a = 2.5 kg, b = 4.5 kg **8.** a = 1.8 litres, b = 3.6 litres **9.** a = 7.4 m, b = 8.6 m

10.

Page 37 ***HWK 2E***

1. (a) false (b) true (c) false (d) true (e) true (f) false
2. (a) 0.01 (b) 0.08 (c) 0.3 **3.** (a) Wesley (b) Alan (c) Kyle (d) 0.11 secs
4. (a) 0.5, 0.518, 0.52 (b) 0.821, 0.83, 0.833 (c) 0.06, 0.094, 0.1, 0.102
(d) 0.32, 0.324, 0.346, 0.35 **5.** 2.99

Page 37 ***HWK 3M***

1. (a) 0.05 (b) 3.2 (c) 0.06 (d) 0.28 (e) 0.018 (f) 3 (g) 0.15
(h) 0.012 (i) 0.042 (j) 0.0008 (k) 0.33 (l) 0.0056 **2.** (a) false
(b) false (c) true **3.** £53.44 **4.** shape B by 0.09 m^2 **5.** 2.36 cm^2

Page 38 ***HWK 3E***

1.

×	0.3	0.05	7	1.2	0.9
0.6	0.18	0.03	4.2	0.72	0.54
0.05	0.015	0.0025	0.35	0.06	0.045
1.1	0.33	0.055	7.7	1.32	0.99
0.8	0.24	0.04	5.6	0.96	0.72
4	1.2	0.2	28	4.8	3.6

2. (a) \$438 (b) \$803 **3.** 62p
4. (a) 0.338 (b) 0.0493 (c) 4.293

Page 39 ***HWK 4M***

1. (a) 80 (b) 430 (c) 6 (d) 700 (e) 20 (f) 2200 (g) 580 (h) 90
2. 24 **3.** (a) false (b) true (c) false (d) false (e) true (f) true
4. (a) 0.01 (b) 0.1 (c) 0.01 (d) 60 (e) 0.1 (f) 0.01

Page 39 ***HWK 4E***

1. (a) 21.3 (b) 1.8 (c) 2.45 (d) 1.93 (e) 12.6 (f) 5.87 (g) 0.169 (h) 13.6
(i) 0.14 (j) 0.832 (k) 7.64 (l) 13.1 **2.** 350 **3.** 85 **4.** 52 hours

Page 40 ***HWK 5M***

GET ME OUT OF HERE NOW

3.2 Using a calculator

Page 40 ***HWK 1M***

1. 8 **2.** 7 **3.** 13 **4.** 5 **5.** 18 **6.** 26 **7.** 99 **8.** 42
9. 13 **10.** 54 **11.** 10 **12.** 50 **13.** 2 **14.** 4 **15.** 2 **16.** (a) true
(b) false (c) false (d) true (e) true (f) false (g) true (h) true (i) true

Page 41 ***HWK 1E***

1. $8 \times 2 + 3 = 19$ **2.** $5 - 3 \times 1 = 2$ **3.** $4 + 18 \div 3 = 10$ **4.** $20 \times 2 - 15 = 25$ **5.** $9 + 16 - 4 = 21$
6. $5 + 9 \times 2 = 23$ **7.** $16 - 14 \div 2 + 1 = 10$ **8.** $10 \div 2 + 3 \times 4 = 17$ **9.** $9 + 4 \times 2 - 3 = 14$
10. (a) $(6 + 3) \div 3 = 3$ (b) $(8 - 2) \times 4 = 24$ (c) $(9 + 1) \div 5 = 2$ (d) $3 \times (6 - 1) = 15$
(e) $28 \div (2 + 5) = 4$ (f) $(8 + 3) \times 3 = 33$ (g) $(5 + 20) \div (3 + 2) = 5$ (h) $30 - (20 - 15) = 25$
(i) $(8 - 1) \times (10 - 3) = 49$ (j) $(10 - 6) \div (1 + 3) = 1$ (k) $(13 + 17) \div 5 = 6$ (l) $15 - (9 + 3) = 3$

Page 41 ***HWK 2M***

1. 33.68 **2.** 16.89 **3.** 10.23 **4.** 5.64 **5.** 5.37 **6.** 8.16 **7.** 9.80 **8.** 0.37
9. 1.71 **10.** 4.34 **11.** 0.67 **12.** 3.14 **13.** 19.63 **14.** 4.44 **15.** 23.83 **16.** 1.73
17. 16.00 **18.** 7.25 **19.** 6.87 **20.** 1.41 **21.** 0.06 **22.** £1133

Page 42 ***HWK 2E***

1. $\frac{18}{35}$ **2.** $\frac{21}{32}$ **3.** $\frac{29}{66}$ **4.** $\frac{9}{35}$ **5.** $\frac{67}{90}$ **6.** $1\frac{23}{40}$ **7.** $2\frac{7}{12}$ **8.** $4\frac{1}{10}$
9. $1\frac{5}{6}$ **10.** $5\frac{1}{24}$ **11.** $3\frac{13}{25}$ **12.** 3 **13.** $3\frac{23}{30}$ hours **14.** $2\frac{1}{6}$ m **15.** shape B by $\frac{1}{2}$ m^2
16. (a) $1\frac{23}{90}$ (b) $\frac{284}{495}$ (c) $1\frac{11}{15}$

Page 42 ***HWK 3M***

1. 33.85 **2.** 2.82 **3.** 57.48 **4.** 3.93 **5.** 1.40 **6.** 1.63 **7.** 1.09
8. 0.07 **9.** 22.05 **10.** 38.56 **11.** 166.14 **12.** 7.08 **13.** 1.05 **14.** 0.93
15. 23.09 **16.** 1.70 **17.** 38.19 **18.** 1.26 **19.** 281.96 **20.** 0.81

Page 43 ***HWK 3E***

1. $\frac{11}{30}$ **2.** 2 **3.** $\frac{19}{30}$ **4.** $\frac{55}{56}$ **5.** $13\frac{1}{12}$ **6.** $3\frac{7}{18}$ **7.** $\frac{67}{140}$ **8.** $\frac{49}{576}$ **9.** $16\frac{1}{11}$

10.

×	$\frac{2}{3}$	$\frac{1}{5}$	$2\frac{1}{4}$	$\frac{5}{6}$	$\frac{1}{2}$
$\frac{3}{4}$	$\frac{1}{2}$	$\frac{3}{20}$	$1\frac{11}{16}$	$\frac{5}{8}$	$\frac{3}{8}$
$1\frac{1}{3}$	$\frac{8}{9}$	$\frac{4}{15}$	3	$1\frac{1}{9}$	$\frac{2}{3}$
$\frac{7}{9}$	$\frac{14}{27}$	$\frac{7}{45}$	$1\frac{3}{4}$	$\frac{35}{54}$	$\frac{7}{18}$
$\frac{4}{5}$	$\frac{8}{15}$	$\frac{4}{25}$	$1\frac{4}{5}$	$\frac{2}{3}$	$\frac{2}{5}$
$\frac{2}{3}$	$\frac{4}{9}$	$\frac{2}{15}$	$1\frac{1}{2}$	$\frac{5}{9}$	$\frac{1}{3}$

Page 44 ***HWK 4M/4E***

1. 4.5 **2.** –32 **3.** 19.2 **4.** 8.5 **5.** –11.1 **6.** 65.9 **7.** 2.8 **8.** –12.2
9. 3.6 **10.** 0.9 **11.** 123.9 **12.** 27.8 **13.** (a) 15.8 (b) –16.2 (c) 4.8
(d) –6.8 **14.** A by 4.4779 **15.** B by 0.385

3.3 Formulas and expressions

Page 44 ***HWK 1M***

1. $a = 17$ **2.** $p = 61$ **3.** $y = 27$ **4.** $m = 4$ **5.** $a = 7$ **6.** $y = 114$ **7.** $m = 161$
8. $a = 6$ **9.** $y = 60$ **10.** $p = 24$ **11.** $p = 58$ **12.** $m = 282$ **13.** A = 60 **14.** A = 1100

Page 45 ***HWK 1E***

1. (a) $x = 39$ (b) $x = 75$ (c) $x = 1$ **2.** (a) $y = 80$ (b) $y = 17.9$ (c) $y = 16.1$
3. A = 192 cm^2 **4.** (a) $m = 48$ (b) $m = 240$ (c) $m = 0$ **5.** (a) $m = 360$ (b) $m = 513$
(c) $m = 873$ **6.** $I = 108$ **7.** $A = 132$

Page 46 ***HWK 2M***

1. (a) 5 (b) 2 (c) 85 **2.** (a) 25 (b) 56 (c) 2 (d) 66 (e) 12 (f) 72
(g) 45 (h) 0 **3.** (a) 1 (b) 25 (c) 51 **4.** $\boxed{x^2 - 1}$, $\boxed{(x-1)^2 + 6}$, $\boxed{\frac{30x}{8}}$, $\boxed{5x - 5}$

Page 47 ***HWK 2E***

1. 2 **2.** 4 **3.** 9 **4.** 5 **5.** 11 **6.** –1 **7.** 32 **8.** 16 **9.** –6 **10.** –3
11. –13 **12.** –6 **13.** 5 **14.** –30 **15.** 5 **16.** –45 **17.** $5n–n$ larger by 2 **18.** (a) false
(b) false (c) true (d) true (e) true (f) false **19.** (a) 14 (b) 76
(c) 16 (d) –45 (e) –1 (f) 28 (g) 198 (h) –18 (i) –115

3.4 Drawing graphs

Page 47 ***HWK 1M***

1. M (4, 2), F (6, 7) **2.** $y = 5$ **3.** C, J **4.** D **5.** G, N **6.** A **7.** M **8.** A
9. F **10.** J **11.** $y = 5$ **12.** $y = 1$

Page 48 ***HWK 1E***

1. C (1, –2), K (–4, 3) **2.** C, L **3.** G, M **4.** D, N, L **5.** K, D, H **6.** $x = 3$ **7.** $y = 3$
8. $y = 4$ **9.** H **10.** N **11.** $x = –3$ **12.** $y = –2$ **13.** $x = 2, y = 5$ **14.** $x = –2$, $y = –2$

Page 49 ***HWK 2M***

1. y: 2, 4, 6, 8, 10, 12 **2.** y: 5, 4, 3, 2, 1, 0 **3.** y: 1, 4, 7, 10, 13, 16 **4.** y: 2, 2.5, 3, 3.5, 4, 4.5, 5

Page 50 ***HWK 2E***

1. lines are parallel, $y = 3x + c$ cuts y-axis at (0, c) **2.** $y = 3x + 4$ **3.** y: 2, 3, 6, 11, 18

Page 50 ***HWK 3M***

1. (a) 18°C (b) 16°C (c) 8°C (d) 08.00 h and 20.00 h (e) 14.00 h and 18.00 h (f) 10°C
2. (b) (i) 9 litres (ii) 8.3 gallons (c) £5.15 **3.** (a) 1400 m (b) 1600 m (c) 1200 m
(d) 2100 m (e) 1100 m (f) 11:00 and 13:00 (g) 2400 m (h) 30 minutes (i) 3 hours

Page 52 ***HWK 3E***

1. $y = x + 4$ **2.** $y = x + 9$ **3.** $y = x – 4$ **4.** $x + y = 9$ **5.** $y = 5x$ **6.** $y = x + 15$ **7.** $y = x – 6$
8. $x + y = 20$ **9.** $y = 8x$ **10.** $y = 2x – 1$ **11.** (a) $y = x + 1$ (b) $y = 3$ (c) $x + y = 3$

3.5 Reflection

Page 54 ***HWK 2M***

2. (e) (–3, 4), (–3, 5), (0, 5), (0, 4) **3.** (d) (6, 3), (6, 5), (5, 5)

Page 55 ***HWK 2E***

1. (f) (2, 4), (2, 5), (–1, 4) **2.** (a) $x = –4$ (b) $x = 1$ (c) $y = –4$ (d) $y = x$
(e) $y = –x$ (f) $y = 0$ (or x-axis) (g) $y = –x$ (h) $x = 5$

UNIT 4

4.1 Describing data

Page 57 ***HWK 1M***

1. (a) 7 (b) 6 (c) 8 (d) 9 **2.** –3 **3.** set B
4. (a) 57 kg (b) 61 kg **5.** £12.50

Page 57 ***HWK 1E***

1. 4 **2.** 5 **3.** bottom half **4.** 5 **5.** 42 **6.** 8 and eg. 1, 12 or 2, 11 or 3, 10 or 4, 9 or 5, 8

Page 58 ***HWK 2M***

1. (a) mean = 21, range = 21 (b) class 8B has the larger mean but the smaller range
2. (a) mean = 58, range = 15 (b) mean = 63, range = 21
(c) group Y has the larger mean and the larger range

Page 59 ***HWK 2E***

1. (a) 108.4 g (b) 110 g **2.** (a) 3.075 (b) 4 **3.** (a) £132.80 (b) £112.80
(c) 'city' families, maybe cheaper local produce for 'village' families, etc.

Page 60 ***HWK 3M/3E***

1.

Stem	Leaf
0	6 8 8
1	2 5 6 6 9
2	0 1 1 2 3 3
3	3 4 4 7 8 8
4	1 3 3 3 7

eg. Key
2|3 means 23

2.

Stem	Leaf
1	0 3 7 7 9
2	1 4 4 8
3	2 7
4	0 8 8
5	2 7 7 8 8 8 9
6	3 4 6 6 7 7 8 9 9

eg. Key
3|7 means 37

3. (a) Halford: range = 38 kg, median = 98 kg Malby: range = 49 kg, median = 107 kg
4. (a) 11.4 cm (b) 6.3 cm

4.2 Rotation and combined transformations

Page 62 ***HWK 1E***

2. (e) (–2, 3), (–6, 3), (–6, 1)

Page 62 ***HWK 2M***

1. (a) (0, 2) (b) (–2, 0) (c) (1, 0) (d) (1, –3)
2. (a) (2, 1) (b) (0, 0) (c) (1, –1) (d) (–1, –1)

Page 63 ***HWK 2E***

1. (e) reflection in the x-axis

2. (e) 1 unit left, 3 units up.

3. (a) reflection in x-axis
(b) 90° clockwise rotation about (0, 0)
(c) 90° anticlockwise rotation about (3, –2)
(d) translation 4 units left, 1 unit down
(e) 90° clockwise rotation about (3, 2)
(f) reflection in x-axis

4.3 Interpreting and sketching real-life graphs

Page 64 ***HWK 1M/1E***

1. (a) 5p (b) 5p (c) 15p (d) 20p (e) 10p (f) 15p (g) 5p
(h) A call costing 15p is between 4 minutes and 6 minutes in length.

3. (a) 11:15 (b) 45 mins (c) 13:30 (d) 40 mph (e) 40 mph

5.

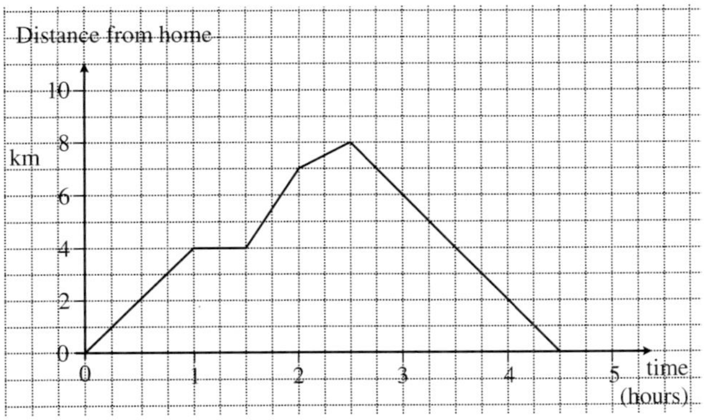

Lucy arrives back home after $4\frac{1}{2}$ hours.

4.4 Brackets and equations

Page 66 ***HWK 1M***

1. false **2.** true **3.** false **4.** false **5.** true **6.** true
7. $6x - 12$ **8.** $8x + 18$ **9.** $15x + 10$ **10.** $mx + nx$ **11.** $ay - by$ **12.** $mn - 4m$
13. $mp + 3p$ **14.** $5b + 5c$ **15.** $xy - bx$ **16.** $ab - ac$ **17.** $3xy + 2x$ **18.** $a^2 + 3a$
19. $x^2 - 7x$ **20.** $4n + n^2$ **21.** $8p - 4$ **22.** (a) $4(3x - \boxed{2})$ (b) $7(\boxed{6} + 3p)$ (c) $3(\boxed{5a} - \boxed{8})$
(d) $\boxed{5}(4 + \boxed{7n})$

Page 66 ***HWK 1E***

1. false **2.** false **3.** true **4.** true **5.** true **6.** true
7. $-3x - 18$ **8.** $-2y + 6$ **9.** $-40 - 5m$ **10.** $-24 + 4n$ **11.** $-10a + 5$ **12.** $-21 - 28p$
13. $-4 - 14x$ **14.** $-24p + 30$ **15.** $-32 + 12m$ **16.** $-9 + 36y$ **17.** $-60 - 50x$ **18.** $-27y + 6$

Page 67 ***HWK 2M***

1. $6x + 22$ **2.** $12x + 26$ **3.** $13x + 40$ **4.** $12x + 48$ **5.** $14x + 32$ **6.** $19x + 41$
7. $27x + 10$ **8.** $44x + 20$ **9.** $29x + 52$ **10.** $12x + 3$ **11.** $11 + 32x$ **12.** $22x + 10$
13. $13x + 17$ **14.** $23x + 72$ **15.** $77x + 62$

Page 67 ***HWK 2E***

1. $5x + 6$ **2.** $13x + 9$ **3.** $20x + 9$ **4.** $7x + 32$ **5.** $27x + 4$ **6.** $9x + 6$
7. $20x + 24$ **8.** $10x + 56$ **9.** $27x + 16$ **10.** $3a + 10b$ **11.** $10a + 18b$ **12.** $16x + 13y$
13. $12m + 3n$ **14.** $7m + 54n$ **15.** $4x + 15y$

Page 68 ***HWK 3M***

1. $m = 16$ **2.** $y = 33$ **3.** $x = 33$ **4.** $n = 27$ **5.** $a = 8$ **6.** $y = 7$
7. $x = 3$ **8.** $y = 9$ **9.** $p = 6$ **10.** $c = 4$ **11.** $m = 12$ **12.** $n = 5$
13. $a = 5$ **14.** $y = 8$ **15.** $x = 1$ **16.** $n = 20$ **17.** $m = 0$ **18.** $y = 3$
19. $m = 11$ **20.** $x = 2$ **21.** $p = 8$

Page 68 ***HWK 3E***

1. $n = 9$ **2.** $x = 5$ **3.** $p = 3$ **4.** $a = 8$ **5.** $y = 7$ **6.** $x = 10$
7. $m = 2$ **8.** $a = 6$ **9.** $y = 15$ **10.** $x = 5$ **11.** $m = 13$ **12.** $p = 20$
13. $n = 10$, side AB = 58 cm **14.** $x = 10$ **15.** $y = 2$ **10.** $n = 45$ **17.** $p = 8$
18. $a = 9$ **19.** $m = 4$

Page 69 ***HWK 4M***

1. 7 **2.** 2 **3.** 8 **4.** 5 **5.** 4 **6.** 9 **7.** 3 **8.** 10
9. 6 **10.** 11 **11.** 1 **12.** 50 **13.** (a) $5(n - 20) = 100$ (b) $n = 40$ **14.** 3
15. 9 **16.** 4 **17.** 20 **18.** 5 **19.** 10

Page 69 ***HWK 4E***

1. (a) $6n + \boxed{12} = 4n + \boxed{20}$, $2n = \boxed{8}$, $n = \boxed{4}$ (b) $12n - \boxed{18} = 8n + \boxed{6}$, $4n = \boxed{24}$, $n = \boxed{6}$
2. 5 **3.** 10 **4.** 2 **5.** 7 **6.** 1 **7.** 14 **8.** 3 **9.** 0 **10.** 8 **11.** 6

Page 70 ***HWK 5M***

1. (a) 18 cm by 6 cm (b) 10.8 cm by 3.6 cm **2.** (a) 15 cm by 3 cm (b) 7 cm by 1.4 cm
3. (a) 9 cm by 7 cm (b) 7.6 cm by 5.6 cm **4.** 8.2 m

Page 70 ***HWK 5E***

1. 4.8 cm **2.** 6.2 cm **3.** (a) 8.6 cm (b) 9.1 cm

Page 71 ***HWK 6M/6E***

1. 14 **2.** 19 **3.** 22 **4.** 31 **5.** 25 **6.** (a) $3x + 26 + 2x + 10 + x = 180$ ($6x + 36 = 180$)
(b) $x = 24°$ (c) 24°, 58°, 98° **7.** (a) $3x + 1 + x + 2 + 3x + 1 + x + 2 = 38$ ($8x + 6 = 38$) (b) $x = 24$
(c) 13 cm, 6 cm **8.** $6n + 3 = 3n + 15$, $n = 4$, Fiona has £27 **9.** $5(7x + 3) = 120$, $x = 3$ **10.** $n = 5$

4.5 Fractions review

Page 72 ***HWK 1M***

1. (a) $\frac{\boxed{16}}{20}$ (b) $\frac{\boxed{15}}{35}$ (c) $\frac{30}{\boxed{54}}$ (d) $\frac{42}{\boxed{48}}$ **2.** $\frac{21}{28}$ **3.** (a) $\frac{11}{15}$ (b) $\frac{5}{24}$
(c) $\frac{1}{12}$ (d) $\frac{3}{10}$ (e) $\frac{17}{20}$ (f) $\frac{41}{63}$ (g) $\frac{11}{30}$ (h) $\frac{19}{40}$ **4.** $\frac{11}{15}$

Page 73 ***HWK 1E***

1. $\frac{2}{21}$ **2.** (a) $1\frac{7}{12}$ (b) $1\frac{9}{10}$ (c) $1\frac{11}{12}$ (d) $3\frac{7}{20}$ (e) $5\frac{1}{6}$ (f) $\frac{5}{6}$
(g) $1\frac{1}{12}$ (h) $2\frac{19}{20}$ (i) $5\frac{3}{20}$ **3.** $\frac{5}{28}$ stone **4.** $\frac{11}{40}$

Page 73 ***HWK 2M***

1. $\frac{5}{6}$ of 12 **2.** £36 **3.** (a) 40 (b) 18 (c) 35 (d) 24
4. trousers £44, shirt £33, coat £132 **5.** 140 minutes **6.** 1280 g **7.** 12

Page 74 ***HWK 2E***

1. (a) $\frac{5}{24}$ (b) $\frac{6}{35}$ (c) $\frac{15}{32}$ (d) $\frac{4}{21}$ (e) $\frac{5}{16}$ (f) $\frac{3}{14}$
(g) $\frac{4}{7}$ (h) $\frac{1}{2}$ (i) 8 (j) 20 (k) 27 (l) $\frac{3}{2} = 1\frac{1}{2}$
2. 2 km **3.** $\frac{3}{20}$ cm^2 **4.** (a) $1\frac{1}{2}$ (b) $5\frac{5}{6}$ (c) 6 **5.** B is larger by $\frac{3}{8}$ cm^2.

Page 74 ***HWK 3M/3E***

1. (a) 0.2 (b) 0.09 (c) 0.45 (d) 0.9 (e) 0.35 (f) 0.25 (g) 0.32
(h) 0.01 **2.** $\frac{13}{20}$ **3.** $\frac{3}{10}$ **4.** (a) 40% (b) 12% (c) 75% (d) 3%
5. 45% **6.** (a) $\frac{1}{50}$ (b) $\frac{7}{100}$ (c) $\frac{21}{25}$ (d) $\frac{8}{25}$ **7.** 0.32, 0.3, 29%, $\frac{7}{25}$, $\frac{1}{4}$

4.6 Handling data

Page 75 ***HWK 1M***

1. (a) Manchester United (b) Manchester United (c) Chelsea
2. (a) 32 (b) 90 (c) 102 **3.** (a) Mark (b) 58 (c) 6
4. (a) £10.75 (b) £6.96 (c) £48

Page 76 ***HWK 1E***

1. (a) group B (b) group B, eg. height values, shape of graph, spread of values
2. (a) 3, 7, 5, 3 (c) 8, 7, 3 (e) Seventeen-year-old times are generally faster because older, etc.

Page 77 ***HWK 2M***

1. (a) one person = 5° (b) milk 160°, dark 150°, white 50°

2.

type of music	Angle
rock	126°
jazz	54°
soul	63°
classical	81°
dance	36°

3.

Capital City	Angle
Paris	90°
Rome	90°
London	120°
Berlin	15°
Madrid	45°

4.

ice-cream	Angle
strawberry	68°
vanilla	50°
chocolate	120°
neopolitan	60°
mint	26°
ginger	36°

Page 78 ***HWK 2E***

1. $x = 48°$

2.

type of vehicle	Angle
car	180°
van	72°
motorbike	54°
lorry	36°
other	18°

3. 'less' – pie charts show proportions not frequencies.

Page 79 ***HWK 3M***

1. (a) 9 (b) 9 (c) true **2.** (a) 8 (b) 5 (c) false

3. (b) In general as the french marks get higher, the german marks get higher.

UNIT 5

5.1 Ratio and proportion

Page 80 ***HWK 1M***

1. (a) 8:3 (b) 1:6 (c) 3:8 (d) 2:3:6 (e) 3:5 (f) 3:2 (g) 7:4:8 (h) 2:5:3
2. 9:7 **3.** 3:1 **4.** 12 **5.** 1:2 **6.** 24 **7.** 20 red, 8 blue **8.** 21 km

Page 80 ***HWK 1E***

1. Nick 12, Beth 20 **2.** red 21 litres, blue 6 litres **3.** (a) £240, £560 (b) £84, £60
(c) £450, £150, £900 **4.** 810 **5.** 80° **6.** (a) £15 (b) 64 kg **7.** 55:25:20
8. 1:3 **9.** 100°

Page 81 ***HWK 2M***

1. £7.02 **2.** £79.20 **3.** £59.15 **4.** 8700 g **5.** 260 **6.** 72 minutes
7. (a) \$58 (b) 27.50 euros (c) Sophie's meal was more expensive by £2 **8.** 49p
9. Dan's deli cheaper by 2p.

Page 82 ***HWK 2E***

1. 8 km **2.** 200 m **3.** 600 m **4.** 130 km **5.** 1.25 km **6.** 6 cm **7.** 60 cm **8.** 32 cm

5.2 Negative numbers review

Page 83 ***HWK 1M***

1. (a) −2 (b) −2 (c) −8 (d) −4 (e) −4 (f) 7 (g) −1 (h) −2
2. (a) false (b) false (c) true (d) true (e) false (f) true **3.** 4
4. (a) $-2 + \boxed{3} = 1$ (b) $-5 - \boxed{-4} = -1$ (c) $\boxed{-4} + (-2) = -6$ (d) $-3 - \boxed{4} = -7$ (e) $\boxed{-1} + (-3) = -4$
(f) $-7 - \boxed{-2} = -5$ **5.** −4 **6.** (a) −4 (b) −4 (c) −8 (d) −6 (e) −2

Page 83 ***HWK 1E***

1. (a) −18 (b) 8 (c) −15 (d) 32 (e) −3 (f) 5 (g) −5 (h) −42 (i) 7
2. 24 **3.** 9 **4.** (a) −2, −10, −5, 20, −2 (b) −4, −16, 2, 18, −6 **5.** −11 **6.** 48
7. (a) $-4 \times \boxed{-10} = 40$ (b) $9 \times \boxed{-4} = -36$ (c) $\boxed{28} \div (-4) = -7$ (d) $\boxed{-5} \times 6 = -30$
(e) $-8 \div \boxed{4} = -2$ (f) $\boxed{-45} \div (-9) = 5$ **8.** (a) −30 (b) −10 (c) −4

5.3 Sequences – the *n*th term

Page 84 ***HWK 1M***

1. (a) $10 \to 30 \to 32$ (b) $20 \to 120 \to 121$, $n \to 6n \to 6n + 1$ (c) $25 \to 100 \to 97$, $n \to 4n \to 4n - 3$
(d) $50 \to 100 \to 106$, $n \to 2n \to 2n + 6$ **2.** (a) $2 \to 10 \to 17$, $3 \to 15 \to 22$, $4 \to 20 \to 27$
(b) $2 \to 12 \to 8$, $3 \to 18 \to 14$, $4 \to 24 \to 20$ (c) $1 \to 9 \to 8$, $2 \to 18 \to 17$, $5 \to 45 \to 44$, $10 \to 90 \to 89$
(d) $1 \to 12 \to 15$, $2 \to 24 \to 27$, $6 \to 72 \to 75$, $8 \to 96 \to 99$

Page 86 ***HWK 1E***

1. (a) $4n$ (b) $4n - 1$ (c) $n - 1$ (d) $3n + 2$

(b) 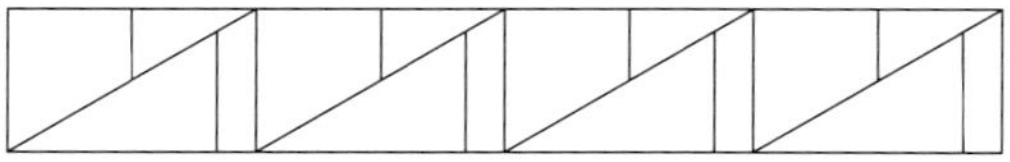

2. (a) 55 (b) false (c) 13 (d) 16 **3.** (a) 10 (e) $3n + 1$

4. (a) $3n + 5$ (b) $2n + 9$ (c) $5n - 4$ (d) $10n - 7$

Page 87 ***HWK 2M***

1. $3n + 2$ **2.** (a) $5n + 3$ (b) $4n - 2$ **3.** $2n + 3$ **4.** (a) $6n - 4$ (b) $3n + 7$ (c) $9n + 4$ (d) $10n - 3$ (e) $5n - 4$

Page 87 ***HWK 2E***

1. $3n + 1$ **2.** $2n + 1$ **3.** $4n + 1$ **4.** $5n + 1$ **6.** $3n + 2$

5.4 Enlargement

Page 89 ***HWK 1M***

7. not an enlargement **8.** × 2 **9.** × 2 **10.** not an enlargement

Page 90 ***HWK 1E***

1. yes **2.** no **4.** $x = 31.5$ cm **5.** 14 cm²

Page 91 ***HWK 2M***

7.

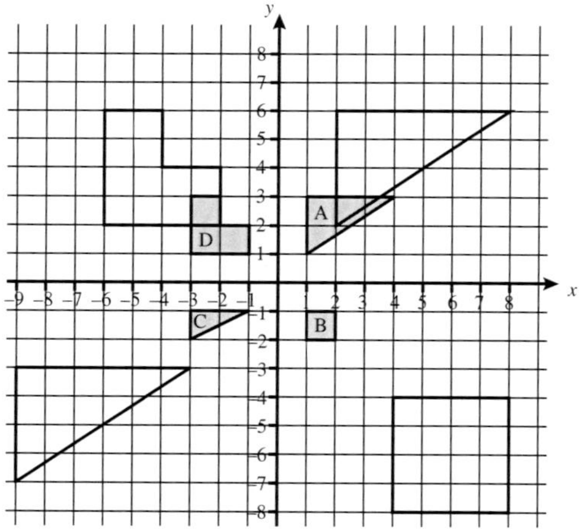

5.5 Congruent shapes, tessellation

Page 93 ***HWK 1M***

1. F, L, M not congruent to any other shape. Congruent pairs are A/N, B/G, C/P, D/I, E/K, H/J

4. Correct

5.6 Drawing graphs review

Page 94 ***HWK 1M***

1. y: 1, 3, 5, 7, 9, 11 **2.** y: 6, 5, 4, 3, 2, 1, 0 **3.** y: 10, 5, 2, 1, 2, 5, 10 **4.** (d) (2, 5) **5.** (2, 6)

Page 95 ***HWK 1E***

1. $y = x + 4$ **2.** $y = x + 7$ **3.** $y = 3 - x$ **4.** $y = 9 - x$ **5.** $y = 5x$ **6.** $y = 5x + 1$

7. (a) $y = x + 2$ (b) $y = 4$ (c) $x = 3$ **8.** (a) $x = 2$ (b) $y = 1$ (c) $y = 6 - x$

5.7 Area Review

Page 96 ***HWK 1M***

1. (a) 24 cm^2 (b) 90 cm^2 (c) 54 cm^2 **2.** 156 m^2 **3.** shape B by 50 cm^2

4. (a) 36 cm^2 (b) 24 cm **5.** (a) 6 cm (b) 12 cm (c) 16 cm

Page 97 ***HWK 1E***

1. (a) 201.1 cm^2 (b) 530.9 cm^2 (c) 1963.5 cm^2 **2.** 549.8 cm^2 **3.** (a) 7.1 cm^2 (b) 77.0 cm^2

4. 30.9 cm^2 **5.** 178.5 cm^2

UNIT 6

6.1 Percentages

Page 98 ***HWK 1M***

1. 85% **2.** 62% **3.** 93% **4.** (a) Jane (55%), Hal (54%), Meg (52%), Tobias (50%)
(b) 2% **5.** 20% **6.** 75% **7.** 80%

Page 98 ***HWK 1E***

1. 53.3% **2.** 77.8% **3.** 10.4% **4.** 52.7% **5.** (a) 21.5% (b) 79.6%
6. (a) Mike (b) 9.8% **7.** (a) 29.9% (b) 43.7% (c) 69.0%

Page 99 ***HWK 2M***

1. (a) false (b) true (c) true (d) true (e) false (f) true
2. £68000 **3.** (a) 30% of £30 **4.** 4% of £500 **5.** (a) £108 (b) £78 (c) £88
6. (a) £2 (b) £1 (c) £3 (d) £7 (e) £21 (f) £10.50

Page 100 ***HWK 2E***

1. (a) £3.42 (b) £54.18 (c) £2496 **2.** (a) 154 (b) 126 **3.** (a) false
(b) true **4.** £7350 **5.** (a) UK 48, Germany 28, Spain 30, New Zealand 28
(b) UK **6.** (a) £23.28 (b) £49.32 (c) £2.08 (d) £138.51
7. (a) £1.35 (b) £5.10 (c) £2.48 (d) £0.67

Page 101 ***HWK 3M***

1. £2000 **2.** (a) £228 (b) £92 (c) £588 **4.** 81 **5.** Amount B by £20
6. £96 **7.** Shirt or skirt **8.** £6210

Page 101 ***HWK 3E***

1. £50.96 **2.** (a) £188.80 (b) £59.15 **3.** Beth by 52p **4.** 637 **5.** £379.20
6. £119.70 **7.** £246.75 **8.** (a) less money (b) £76.32

6.2 Probability

Page 102 ***HWK 1M***

1. (a) $\frac{1}{6}$ (b) $\frac{1}{3}$ (c) $\frac{1}{3}$ (d) $\frac{1}{2}$ **2.** 0.03 **3.** (a) $\frac{1}{6}$ (b) $\frac{2}{3}$ (c) $\frac{1}{3}$ **4.** (a) $\frac{1}{9}$ (b) $\frac{7}{9}$
(c) $\frac{1}{3}$ **5.** 53% **6.** (a) $\frac{1}{4}$ (b) $\frac{3}{4}$ **7.** (a) $\frac{1}{4}$ (b) $\frac{5}{8}$

Page 103 ***HWK 1E***

1. (a) $\frac{2}{5}$ (b) $\frac{4}{7}$ **2.** (a) $\frac{1}{2}$ (b) $\frac{3}{8}$ (c) $\frac{4}{9}$ **3.** (a) $\frac{2}{7}$ (b) eg. 5 (c) eg. 10, 15
4. (a) $\frac{3}{7}$ (b) $\frac{1}{3}$ **5.** (a) $\frac{1}{4}$ (b) $\frac{1}{2}$

Page 104 ***HWE 2E***

1. (a) $\frac{23}{80}$ (b) $\frac{34}{80} = \frac{17}{40}$

2. (a) The dice hardly landed on a '3'.

(b) 20 times for each number.

(c) No

3. (a) $\frac{28}{50} = \frac{14}{25}$ (b) $\frac{57}{130}$ (c) $\frac{85}{180} = \frac{17}{36}$

4. Expect to get 25 tails.

Page 105 ***HWK 3M/3E***

1. (a)

pizza	coke
pizza	lemonade
pasta	coke
pasta	lemonade

(b) $\frac{1}{4}$

2. (a)

boy	boy
boy	girl
girl	boy
girl	girl

(b) $\frac{1}{4}$

3. (a) (H 1), (H 2), (H 3), (H 4), (H 5), (H 6)
(T 1), (T 2), (T 3), (T 4), (T 5), (T 6)

(b) $\frac{3}{12} = \frac{1}{4}$

4. (a) (1, 1), (1, 2), (1, 3), (1, 4), (1, 5), (1, 6)
(2, 1), (2, 2), (2, 3), (2, 4), (2, 5), (2, 6)
(3, 1), (3, 2), (3, 3), (3, 4), (3, 5), (3, 6)

(b) $\frac{6}{18} = \frac{1}{3}$

5. (a) (1, 1), (2, 1), (3, 1), (4, 1), (5, 1), (6, 1)
(1, 2), (2, 2), (3, 2), (4, 2), (5, 2), (6, 2)
(1, 3), (2, 3), (3, 3), (4, 3), (5 ,3), (6, 3)
(1, 4), (2, 4), (3, 4), (4, 4), (5, 4), (6, 4)
(1, 5), (2, 5), (3, 5), (4, 5), (5, 5), (6, 5)
(1, 6), (2, 6), (3, 6), (4, 6), (5, 6), (6, 6)

(b) $\frac{6}{36} = \frac{1}{6}$

6. $\frac{1}{8}$

6.3 Measures

Page 106 ***HWK 1M***

1. 7.4 km **2.** 4 cm **3.** 0.35 m **4.** 2500 g **5.** 4.5 litres **6.** 6800 m **7.** 0.32 kg
8. 6500 kg **9.** 238 cm **10.** 4200 m **11.** 3400 ml **12.** 0.4 tonnes **13.** 1830 g **14.** 3 times
15. 1 tonne block **16.** £63000 **17.** 1488 kg **18.** 17

Page 107 ***HWK 1E***

1. (a) 16 miles (b) 5 feet (c) 20 feet (d) 7 pounds (e) 5 gallons (f) 12 pints
(g) 7 stones (h) 60 km (i) 90 litres (j) 4 pounds **2.** 4 inches **3.** 1.5 litres
4. (a) true (b) true (c) false **5.** the woman by 3 kg

6.4 Algebra Review

Page 108 ***HWK 1M***

1. 6 **2.** 45 **3.** 17 **4.** 87 **5.** 10 **6.** 81 **7.** 56 **8.** 12 **9.** 4 **10.** 60
11. 90 **12.** 8 **13.** $8m + 4n$ **14.** $9p + 3q$ **15.** $5a + 3b + 3$
16. $15 + x$ **17.** $y - w - 5$ **18.** $ac + bd$ **19.** $y = 27$ **20.** $\frac{1}{2}m - 16$

Page 108 ***HWK 1E***

1. (a) false (b) false (c) true (d) true (e) false (f) true **2.** $m - n - 9$
3. $pq - mn$ **4.** 5 **5.** 32 **6.** 9 **7.** 60 **8.** 42 **9.** 27
10. 136 **11.** $3m - 2$ **12.** $3n - 4$, $2n + 5$

Page 109 ***HWK 2M***

1. $5n + 20$ **2.** $3n + 3$ **3.** $6n - 42$ **4.** $9n - 6$ **5.** $24n + 30$
6. $36 + 27n$ **7.** $14n + 35$ **8.** $12 + 4n$ **9.** $15 - 10n$ **10.** $mn + 4n$
11. $ab + ac$ **12.** $ny - 9n$ **13.** $3x + xy$ **14.** $w^2 - 2w$ **15.** $30m + 6$
16. $5m + m^2$ **17.** $n^2 - np$ **18.** $ab + a^2$

Page 109 ***HWK 2E***

1. $6n + 26$ **2.** $10a + 46$ **3.** $11m + 34$ **4.** $14y + 17$
5. $32a + 66$ **6.** $38m + 54$ **7.** $29p + 75$ **8.** $54x + 78$
9. $9a + 9$ **10.** $2m + 14$ **11.** $n + 27$ **12.** $2x + 19$
13. $16y + 60$ **14.** $11w + 3$ **15.** $38n + 14$ **16.** $23a + 52$

Page 110 ***HWK 3M***

1. 22 **2.** 5 **3.** 4 **4.** 8 **5.** 7 **6.** 10 **7.** 5 **8.** 12 **9.** 4 **10.** 13 **11.** 5
12. 12 **13.** 8 **14.** 11 **15.** 4 **16.** 4 **17.** 5 **18.** 4 **19.** 2 **20.** c = 8

Page 110 ***HWK 3E***

1. 6 **2.** 4 **3.** 10 **4.** 9 **5.** 15 **6.** 7 **7.** (a) 8 (b) 74 cm **8.** 7
9. 5 **10.** 9 **11.** 2 **12.** 4 **13.** 10 **14.** 8 **15.** 12

6.5 3–D Objects

Page 111 ***HWK 2M***

1.

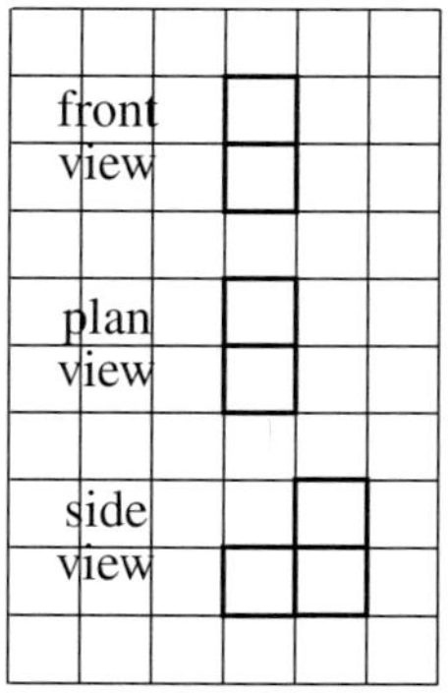

2.

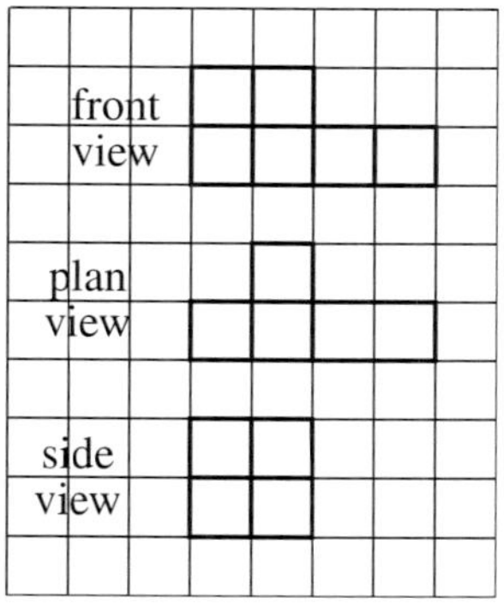

3.

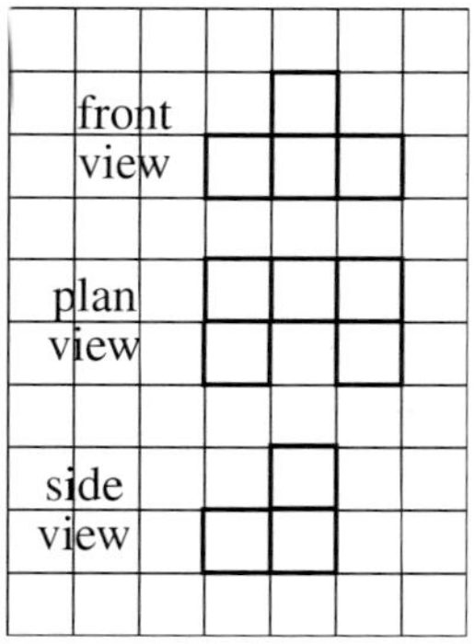

4.

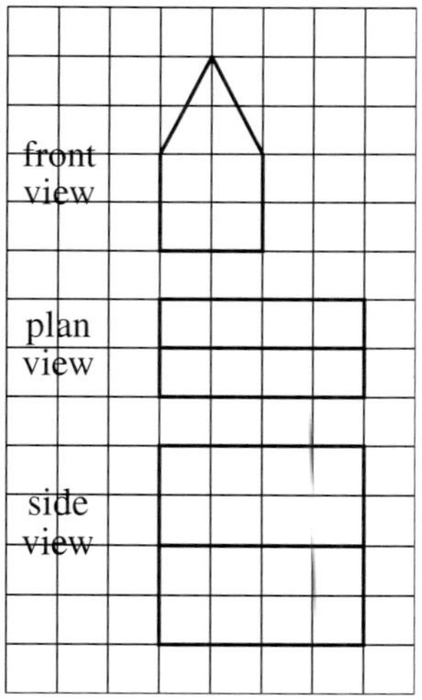

5.

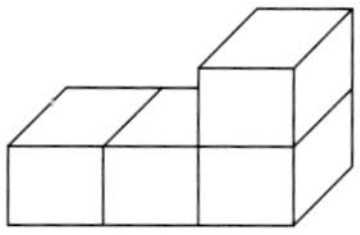

6.

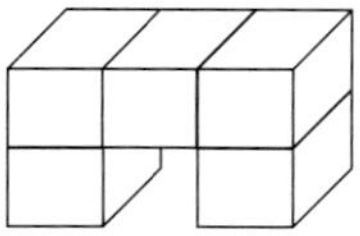

7.

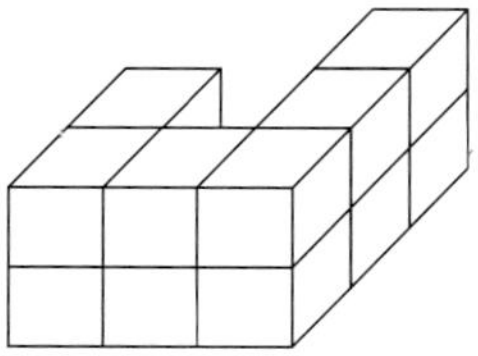

8.

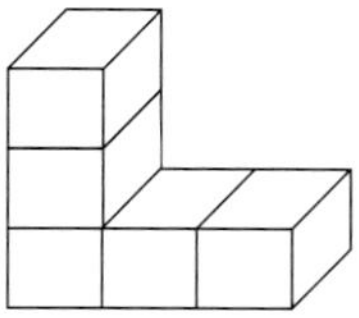

6.6 Bearings and scale drawing

Page 112 ***HWK 1M***

1. A 256°, B 043°, C 304°, D 156°, E 060°, F 265°, G 121°, H 341°, I 212°, J 101°

2. (a) 140° (b) 320° **3.** (a) 065° (b) 160° (c) 105° (d) 285°

Page 113 ***HWK 1E***

1. (a) 334° (b) 096° (c) 150° (d) 229° (e) 063° (f) 270° (g) 180°

3. (a) 050° (b) 282° (c) 211° (d) 078°

Page 114 ***HWK 2M***

1. 50 m **2.** 13.4 m **3.** 21.6 m

Page 115 ***HWK 2E***

1. 9.6 km **2.** 9.2 km **3.** 9.4 km **4.** 10.0 km **5.** 7.9 km **6.** 227°

6.7 Decimals Review

Page 116 ***HWK 1M***

1. (a) false (b) true (c) true (d) false **2.** Kelly, Bryony, Cameron, Sarwan, Lucy

3. 2.29 **4.** 31.4 – 8.18 **5.** 4.15 **6.** 3.675 **7.** 8.7 **8.** 8.62

Page 116 ***HWK 1E***

1. (a) false (b) false (c) false (d) true **2.** 10 **3.** 0.07, 0.056, 2.8, 14

4. £104.30 **5.** (a) 0.0021 (b) 4.8 (c) 3.12 **6.** £1.86 **7.** (a) $0.7 \times \boxed{1.6} = 1.12$

(b) $\boxed{0.324} \div 0.09 = 3.6$ **8.** A by 0.42 cm^2

6.8 Volume

Page 117 ***HWK 1M***

1. (a) 22 cm^3 (b) 8 cm^3 (c) 30 cm^3 (d) 22 cm^3 **2.** (a) 128 m^3

(b) 1200 cm^3 (c) 216 mm^3 **3.** P by 2 cm^3 **4.** 1176 cm^3

Page 118 ***HWK 1E***

1. 25600 mm^3 **2.** 0.43 m^3 **3.** 26.25 cm^3 **4.** 6600 m^3 **5.** (a) 5 cm

(b) 6 cm (c) 7 cm **6.** (b) 140 cm^3 (c) 166 cm^2 **7.** 150 minutes

8. 13824 **9.** 18*abc* **10.** 24*mnp*